Shock and Awe
Basic Training

Vickey Joins the Military Series

Book 2

Shock and Awe
Basic Training

Book Two

Table of Contents

Basic Training

The Initial Encounter!

It was finally time to leave the reception center and head to the basic training station. The personnel at the reception center had attempted to describe the upcoming experience, but their words couldn't do it justice; to truly believe what they were telling us, you had to experience it yourself.

The transportation that would take me and forty-nine others to the basic training station arrived; it was a literal cattle trailer. As the cattle trailer pulled up, the sergeant, speaking in a calm, quiet voice, instructed us to load our bags into the undercarriage.

After loading our bags, we began to file onto the transport, squeezing in as tightly as possible. The goal was for all fifty of us to fit into the single vehicle, though it was clear this was a tight squeeze beyond what we had intended.

As we packed ourselves in, a voice called out, "Make your buddy smile," encouraging a bit of camaraderie as we settled into the cramped space.

“Everyone squeeze in,” the sergeant yelled. It was no longer a question of whether we could all fit; it was already happening.

Thankfully, the ride to our new living quarters wasn’t long, because it was quite uncomfortable.

When we arrived at the basic training site, life as I knew it changed dramatically, and it was over, really over.

A sergeant wearing unusual headgear barked at us, “You maggots, get out here now!”

“Oh, my goodness, what is happening right now? What is with all the yelling and rushing?” I wondered, overwhelmed by the sudden shift in atmosphere.

“Who are these loudmouth individuals?” I thought to myself as there was yelling instructing us to disembark from the cattle trailer. It was unclear why the sergeant was yelling. We hadn’t done anything wrong. We were simply following orders.

As we filed out of the trailer, we were expected to form an orderly gathering in smaller groups. However, it seemed we weren’t meeting the sergeant’s expectations, as he continued to yell, clearly dissatisfied with our efforts.

“Obviously you maggots lack discipline,” he yelled.

"But you know, I can fix that. DROP!"

"Oh, my goodness, what is happening right now" I thought once again, but that was all the time I had to process the situation. Almost immediately, I found myself on the ground, struggling to do a push-up among a series of other grueling exercises that could make anyone cry.

We crawled on the ground, we rolled over, we held our legs six inches off the ground, and we ladder climbed.

The yelling never stopped; it filled the air as we, the trainees, scrambled to comply with the relentless stream of commands. We were pushed to our limits, attempting each exercise as demanded. Amidst the chaos of shouted instructions and physical exertion, I couldn't help but wonder what I had gotten myself into.

"Why did I leave a quiet, promising career for this?" I wondered, on the verge of tears. The realization hit me hard: I was in the Army now.

We began to march, or more accurately, trudge along while carrying our heavy duffle bags loaded with military gear and our civilian luggage.

With a duffle bag strapped to my front, another on my back, and my civilian luggage dragging behind, I desperately wished for the day's agony to end.

"Oh, my goodness, this is excruciating, please make it stop. Will this ever end?" These thoughts raced through my mind. Finally, the sergeants allowed us to put everything down.

The next few days were all about settling in and learning some basic tasks before the official beginning of the eight-week basic training session.

We would later understand that the main goal of training was to stop thinking individually and start working as a team. Basic training emphasized doing everything the same—from wearing uniforms to helping each other—to avoid getting in trouble together. But this wasn't clear to us right away.

Red Phase

Torture Overload

Each of us was assigned one large locker to keep all our belongings in and a bunk to sleep in.

Welcome to basic training life.

Rooms had to be spotless at all times. Bunks were to be made with 45-degree folds at the corners, shoes lined up neatly at the foot of the bed, and lockers organized precisely. Additionally, the rooms needed to be swept, mopped, and buffed every day.

We were caught in a never-ending cycle of chores, and basic training hadn't even officially started yet.

The lesson was clear: if we followed orders, we'd seldom face any problems.

But this was a hard lesson to get across to a bunch of individuals. The days echoed with constant yelling. By the second full day of basic training, I was deeply aware that we were at the start of some truly challenging times ahead.

"You people want to be late to my formation, down," yelled the drill sergeant, "drop and eat dirt."

"Beat your faces." Don't you dare put your knees on that ground, you better hold yourself up until I tell you otherwise."

The torture continued day after day, and it was the worst thing I had ever endured. The only break from the physical torture came during briefings.

With the entire focus of basic training being to transform individuals into a team, the red phase served as the catalyst to that transformation. We started to see that the hardships were shared by all, helping us embrace the idea of being "together as one."

It became clear that no individual was singled out; rather, we were all subjected to equal levels of hardship. That realization fostered a sense of solidarity among us. When one trainee was required to drop and do push-ups, the whole group would join, marking the early stages of team development.

The red phase was dedicated to transforming us into a cohesive team, filled with rigorous physical training and basic soldiering skills. Our days began at "zero-dark-thirty" and included elements like intense exercise, several miles of running, inoculations, briefings, and a continuous push to break us down and build us up as a unit.

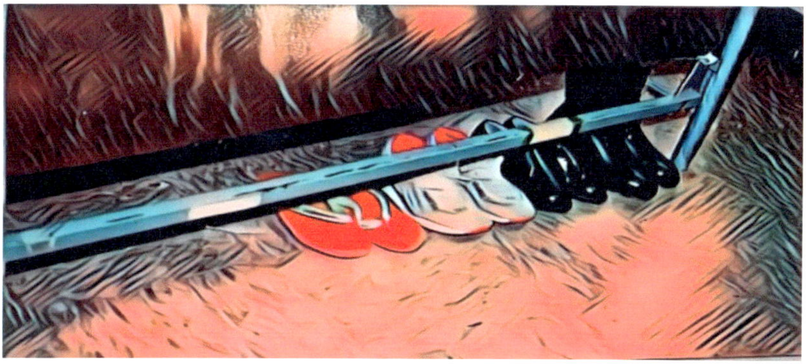

Everything had to be dress-right-dress—perfectly aligned and orderly!

My first official task in basic training was to complete an initial physical training (PT) test. Unfortunately, I didn't perform as well as I had hoped. I managed only seven correct push-ups and thirty-six correct sit-ups.

Later that day, I reflected, "If I had pushed a little harder, I could have done 10 push-ups and 40 sit-ups." I also needed to complete my two-mile run in 18:54 or less; I finished in 19:10, which was a failing time. Despite that, my time was among the top five best in my group.

I was not ready for the extreme weather conditions I faced in basic training. The temperatures were frightfully blistering. Coming from the south, the bitterly cold weather seeped into my bones. Just a week into training, it felt as though I had been there an eternity. I couldn't see an end to the torture.

However, as the days started to blur into a repetitive cycle, much like Groundhog Day, and the red phase—the toughest of all—was coming to an end, I realized I hadn't faced anything too extremely difficult or spirit-breaking.

As the red phase started to wind down, the key objectives were to pass the Phase I test and a second PT test.

I excelled in the Phase I test, which covered basic military knowledge, and I improved my push-ups from seven to twelve and sit-ups from thirty-six to fifty. However, my time for the two-mile run increased, slipping from 19:10 to 20:47; my time was going in the wrong direction.

The Chaplain told us, "If you all can make it through the red phase, the rest will be a lot easier." Successfully completing the red phase, which is the toughest part of basic training, meant that the subsequent white and blue phases would be more manageable, with each phase being slightly less challenging than the one before.

The day finally arrived to switch over from red to white phase. However, that did not mean everything would just instantly be easier.

We stepped into the white phase with cautious optimism.

White Phase

Basic Soldiering

The white phase began with the receipt of a M16A1 rifle and field gear. I entered the phase with definite shock and awe.

"This is my weapon," became a phrase we had to repeat over and over.

"Listen up, this is not a gun," the drill sergeant yelled, emphasizing the significance of the equipment we were handling.

"You will NEVER call it a gun. It is a weapon."

If the M16A1 was ever referred to as a gun, it was always followed by strenuous activity.

I had never handled or fired any type of weapon in my life, and suddenly I was required to learn how to operate and qualify with an M16A1 semi-automatic rifle at a shooting range. The experience was both exhilarating and frightening simultaneously.

And although qualifying with the weapon was necessary to graduate from basic training, it was not the first hurdle we had to successfully maneuver in this phase.

While the focus was on field training and basic rifle marksmanship (BRM), the phase actually started at the gas chamber. That was where we had to learn how to protect ourselves against chemical, biological, or nuclear nerve agents—a critical and daunting part of our training.

"Listen up," the drill sergeant yelled out.

"Twenty of you will enter the chamber at a time, and in rows of five, you will exit."

On the surface, the instructions for the gas chamber didn't seem too daunting. The plan was for twenty of us to enter the chamber, and then exit five at a time. It appeared to be a straightforward task that wouldn't be too difficult to manage.

Twenty of us entered the chamber, masks on, and arranged ourselves into rows. Standing there, fully masked and organized, awaiting further instructions, there was no hint of the drama and shock that was about to unfold.

Now organized inside the chamber, our row was instructed to remove our masks on command. After removing the mask, we were to state our name, rank, and social security number. Again, I thought, okay, this seems doable.

We had NO idea how extremely difficult the task would be!

The first row was instructed to remove their masks and, one by one, state the required information. From the moment the first row removed their mask, they met with SHOCK!!

The moment the seal of their masks broke and they were exposed to the gas, the chemicals in the chamber overwhelmed the trainees. They began violently choking, with secretions streaming from their eyes and nose. There was extreme difficulty trying to speak, to provide the three pieces of information required to submit.

Watching the others before me endure such intense distress, I hurriedly tried to devise a plan for how I might manage this seemingly impossible task. However, I quickly realized that the effectiveness of my individual performance was irrelevant; no one from our row could leave the chamber until every one of our row-mates had completed the task. That added an extra layer of urgency and collective responsibility to the challenge.

As I watched the horror play out, time was growing near for my row to attempt the feat.

The time had come, and my row had geared up to remove the mask. The difficulty level had become painfully clear.

“Remove your mask,” the drill sergeant yelled through his own mask.

Here, I was hoping that everyone in my row would remove their mask at the same time and attempted to formulate in their mind a strategy to accomplish the task. There was no amount of strategizing that could have helped us through the chamber ordeal. We had no choice but to endure it, hoping that our time unmasked in the chamber would be as brief as possible.

With our masks off, we endured the same fate as those before us—choking and experiencing an abundance of secretions. Despite our desperate desire for the ordeal to end, it felt like an eternity before it would be over.

I managed to provide my name, rank, and the start of my social security number, which I thought was pretty impressive given my condition. As the drill sergeant went down my row seeking the information from other row-mates, I was pulling my hair, figuratively, and waving my hands in the air, jumping up and down and trying to head for the door.

But of course, I couldn't leave until instructed to do so. Instead, I was told to get back in line and wait, enduring yet another challenge that was completely new to me.

Finally, the torture came to an end, and my row was allowed to depart the chamber. However, we couldn't run out; doing so would mean having to stay longer. Fortunately, none of us in my group had to reenter the chamber.

Once outside, we were instructed to run around two light poles and flap our arms—a command that sounded ridiculous at first. However, once we went through the experience and ran, we understood the purpose behind it.

The fresh air helped clear the gas from our bodies, and the arm waving helped expedite the process. Once the group was stationary, the event did not seem that painstaking and we all joked about what we did while we were in the chamber, but that was an experience I had no desire to endure again.

The drill sergeants wasted no time in informing us that the simulated event we had just experienced was nothing compared to enduring an actual chemical, biological, or nuclear gas event. They emphasized that a real event would be four times as deadly. That truly put the severity of the situation into perspective. Wow!!

We were in the chamber, unmasked for only 15 seconds, yet it felt like an eternity. The instructors explained that any exposure longer than a minute to actual nerve gas would be deadly, or we would suffer from serious nerve damage. It was a truly eye-opening experience!!

The chamber marked merely the start of the soldierly experiences that we civilians would undergo. It was time to put our freshly acquired soldier skills to use in a field setting.

"No seriously what exactly is this to entail," I thought to myself.

"I'm a city girl, and I am not going to fare well in the outdoors." The unknowns just kept coming.

The field week kicked off with my fireguard duty at 1:00 am, lasting an hour. Later that morning, we started the day as usual with PT and other tasks before heading out to the field. There, we would dine from mess kits, no showers, and live simply in the wilderness.

"What, we will be sleeping in a tent?" Oh, and we, the city folk, had to assemble the tents.

There were no beds; instead, we slept in sleeping bags on the ground—something I had never considered before entering basic training. And here's another surprise: no hot meals. We subsisted on packaged meals, Meals Ready to Eat (MREs), which were our daily fare.

Will the endless surprises come to a halt? No, not for a while.

Here we were at a pivotal point in training. If we could make it through two more weeks, we would be at the halfway point of basic training, having completed four out of the eight weeks.

We arrived at our campsite just before dark.

Unbeknownst to us trainees, each of us had been issued only a half tent with our field gear—a fact we only realized upon arriving at the campsite and receiving instructions to assemble our tents.

As we trekked to our site, each trainee, carrying a shelter-half, was paired up. The buddy teams had to marry their shelter halves together to create a full sleeping tent.

"This is a really small area," I immediately thought to myself. And each buddy team had to place rucksacks, large duffle bags, sleeping bags and pads, weapons, other gear, and ourselves into the small shelter tents. Somehow, it all seemed to fit.

But the most shocking element to the field exercise was the outdoor bathroom and shower facilities. Let's just say, there was very little privacy at these personal facilities!

As the days in the field wore on, filled with four-to-five-mile marches, rifle marksmanship, grenade training, and land navigation exercises, the day finally came for another milestone: the 12-mile ruck march.

The day started early, like all the others, and with a 30-pound rucksack on my back, the journey seemed endless. Just when I thought I couldn't go any further, I pushed a little harder and discovered new strength. I always knew I had perseverance, but during basic training, I truly tested its limits.

The long, hard, arduous march ended at our next campsite. Upon arrival, we were instructed to dig foxholes—protective areas meant to serve as cover during enemy engagements. That was where we would also pull guard duty.

Already exhausted, we began digging immediately, creating foxholes armpit-deep to the tallest person among us. And we still had to assemble our tents.

I immediately thought, "I am going to die out here. Seriously, they expect us to dig a four-foot hole and set up our tents after getting up early and trekking twelve miles?"

It was cold and wet, and I was truly miserable pulling guard duty. I made a colossal mistake in that situation, I pulled duty and slept in my full, cold, wet gear and I had no way of warming up.

"I need this night to be over, make it stop," I thought as I sat every other hour in the foxhole in the rain, in freezing temperatures.

Although long, the night eventually ended, and as with all the tasks to date, upon reflecting, I remembered thinking, "It really wasn't that bad." Of course, that thought came after surviving the cold, wet night. But I also distinctly remembered thinking, "I'm glad that's over."

We had gone days without a shower and our clothes were filthy; that was how we entered the next event of the field exercise.

"It's dark, I can't really see what is happening," I was worried and scared of the night event. We went through a night course with bullet rounds fired at us as we low crawled, high crawled, and did three to five-second rushes through the course.

"I am hating this entire week." Although we had the opportunity to change clothes—not to take a shower—the sense of cleanliness was short-lived. With the next event, we were dirty once again.

Dirty, tired, and hurting, the brutality of the week pressed on. The next course involved practicing defensive crawls. We drilled high crawls, low crawls, and firing our weapons for protection at the end of each sequence. Exhaustion was beginning to take its toll, but I persisted, completing each drill. As the drills extended into the night, we transitioned into night training, which made the maneuvers a bit more complicated.

After completing the drills and what felt like hours later, we prepared to shoot "night fire" on the rifle range.

"It's raining, I'm cold, I'm tired, I smell bad, and I'm ready to get back to the barracks," were the thoughts running through my mind. I was miserable but returning to the warmth of the barracks wouldn't come until we had finished the task at the night rifle range.

"Hallelujah!!!!" We get back to the barracks and we get a chance to stay up until midnight to shower, wash our hair, and do anything else we needed to do. It was great being back at the barracks. I never would have thought that I would be so excited to return to a place that I had dreaded before field week. It's amazing how experiences can shift our perspectives and make us appreciate things we once took for granted.

Blue Phase

Prepping For Transition

Entering the blue phase filled me with glee and a sense that my time in basic training was nearing its end. As we trainees recovered from the long, wet, cold field training, we had to gear up for our final exit events which involved an exit PT exam to name the biggest hurdle. Knowing that PT was the leading cause of failure, we began some intense, hellacious training.

But with basic training smack dab in the middle of Christmas time, there were some light moments. We were allowed to attend a USO concert, an informal Christmas program, participate in some intramural sports, and even had some time off to wander around the post on our own.

“Ah, it is Christmas Eve and I just feel the stress leaving my body,” I remembered.

On that day we had breakfast, had our Class A’s or dress uniform inspected, and went to a Christmas party.

“It is almost over,” I kept thinking, “It is almost over.”

The day ended with us being released to get away, do normal things, and be with family.

The first thing we did was head out to catch the shuttle to get away and begin celebrating freedom. Many of us had family members and hotel rooms waiting for us.

First things first, ditch the uniform and head to the restaurant for a “real” meal where we could take our time eating and talking. It was so nice being a civilian, if only for a brief time. Civilians again, yeah buddy.

The End of Basic Training

The break from soldiering was brief. We said goodbye to family and friends, and then we had to return to training to complete our course.

"Oh, so happy for a brief moment," I thought, but that moment was well worth it, if for no other reason but to maintain sanity.

As we returned to training, we were all still happy and reeling from our much-needed time off, but we had to return to our reality, the reality that we had to complete training before we could move on.

We reentered training with smiles and armed with the knowledge that our time in basic was drawing to a close. We began preparing for our end-of-cycle tests, both written and physical.

On the early morning of the final PT test, I was extremely anxious. That morning was unlike any other because failing to perform well in that session meant staying in basic training, not graduating, and not advancing to the next level of job-related training.

As we all lined up in about six rows, the time had finally come for the final PT test, which consisted of two minutes of push-ups, two minutes of sit-ups, and a timed two-mile run. I was feeling the pressure.

I exceeded my wildest expectations for my results. I managed twenty-two push-ups, fifty-five sit-ups, and I completed the two-mile run in 16:34. That was remarkable for someone my age competing with 18 and 20-year-old women. I scored 68, 72, and 97 points respectively, totaling 237 out of a possible 300 points. With adrenaline racing through my veins, I felt highly motivated to tackle the next end-of-cycle test: the written exam.

On the written exam, I scored a near-perfect score, nineteen out of a possible twenty. It was now OVER; I had passed the key requirements to move on from basic training. And to top basic training off, I got on the scale and found out I had lost fourteen pounds during training. Oh, happy days, on to graduation!!

Leading up to the final days of basic training, we cleaned and turned in our military-issued equipment, processed out, signed papers, and practiced graduation ceremonies.

During eight weeks of basic training, I had endured a gas chamber, a cold/wet field exercise with bullets flying overhead as I low crawled through dirt and mud, threw grenades, trekked more than 12 miles with a 30-pound rucksack on my back that left my feet blistered, and endless yelling and exercise, leaving with a mindset, "Will this ever end?"

The sole focus of the training was to transform us into a "we" and remove "I." Accomplished, because we had to be a "we" to succeed at the given tasks.

The big day finally arrived. We graduated, received our plane tickets, and headed to our next destination: advanced individual training.

It was a bittersweet moment as some of the people who started the basic training journey with me didn't graduate due to failing their final PT test. It was sad to leave behind those I had bonded with. Yet, it was also a day of joy because I was moving on to the next phase of my training.

I caught a bus to the airport, flew for hours, and finally arrived at my advanced individual training location in the desert. That was the last leg of training before I became a full-fledged soldier.

Not a soldier yet!

Made in the USA
Las Vegas, NV
07 November 2026

57669948R00019